What Others are Saying About
HEAT + PRESSURE

"In *HEAT + PRESSURE*, Ben Weakley unpacks memory, pain, loss, and a generational inheritance of war and conflict. These verses attend to the physical of this world with a clear-eyed examination of the self—all in the service of gleaning insight into the messy and layered humanity within us all. While *HEAT + PRESSURE* traverses the interior and exterior terrains of war, damage, and destruction, these poems are watermarked with tenderness and care—a sure sign that these are poems rooted in the word *love*."

—Brian Turner, author of *Here, Bullet* and *Phantom Noise*

❖ ❖ ❖

"In tight, considered lines, Ben Weakley reminds us that war—past and present—is a human experience, a rite-of-passage turned horror show that only the few endure, its impact echoing through time [...] Like the blasts he writes about, you will feel these poems reverberate in your being. This is what it's like to go to war."

—Colin D. Halloran, author of *Shortly Thereafter*,
American Etiquette, and *Icarian Flux*

❖ ❖ ❖

"Ben Weakley's keen eye for details both physical and emotional have enabled him to pen a searing, visceral, and lyrical portrait of the experience of the Global War on Terror. More than that, he explores themes of fatherhood and early middle-age in modern America through the lens of the war's aftermath, illustrating the longevity of its impact with skill, nuance, and heart. *HEAT + PRESSURE* is an unparalleled collection of poetry that offers its readers the full breadth of the contemporary veteran's experience."

—David P. Ervin, author of *Leaving the Wire: An Infantryman's Iraq*

❖ ❖ ❖

"Weakley's poems expertly track the jagged and jarring arc of war. Beginning with boys who play in 'caves of their imagination' and pray to the 'god of manhood.' Duty and fate turns them into soldiers who 'seek enlightenment in dust clouds' and experience the pain and pleasure of the fight. [...] Veterans of all eras will recognize the scents, scenes, and scars that Weakley describes."

—F.S. Blake, author of *Terminal Leave*,
Above the Gold Fields, and *The Few Drops Known*

❖ ❖ ❖

"Weakley's book is framed by the elements needed to turn C-4 explosive from a block of 'Play-Doh' into something that rends metal and incinerates bodies: *HEAT + PRESSURE*. His poems crackle and stay with the reader for days. A great read!"

—Christopher Lyke,
author of *The Chicago East India Company*

❖ ❖ ❖

"*HEAT + PRESSURE* is aptly named. These short poems are combustible, a Molotov cocktail of bearing witness to the inhumanity of war sloshing together with the snippets of how you can end up on the wrong end of a rifle, which is any end of a rifle. This debut collection takes its rightful place on the list of must-read books of war poetry."

—Martin Ott, author of *Lessons in Camouflage*
and *The Interrogator's Notebook*

HEAT +
PRESSURE

Poems from War

Ben Weakley
Middle West Press LLC
Johnston, Iowa

❖ ❖ ❖

Poetry / U.S. Wars in Iraq & Afghanistan / Military Life

HEAT + PRESSURE: Poems from War
by Ben Weakley

ISBN (print): 978-1-953665-14-0
ISBN (e-book): 978-1-953665-15-7
Library of Congress Control Number: 2022942478

❖ ❖ ❖

Middle West Press LLC
P.O. Box 1153
Johnston, Iowa 50131-9420
www.middlewestpress.com

❖ ❖ ❖

*Special thanks to James Burns of Denver, Colorado!
Your patronage helps publish great military-themed writing!*
www.aimingcircle.com

For Stefanie, Abby & Jack

CONTENTS

Debris

Fragmentation

Heat

When we were boys

every grandfather contained stories
from The War, told in whispers from other rooms.

Their trophies preserved eternities
fixed between the quick and the dead.

Framed inside a shadow box on the wall,
my best friend's grandfather kept

a utility sleeve and brass casings
from two Japanese bullets that sailed

close enough to rip the fabric but miss
his young muscle and bone.

Another friend's grandfather trudged
across Burmese jungles and swamps

picking off Japanese officers,
disappearing into mist.

Long after the war, he wouldn't lie
down in a hammock—in his dreams

the Emperor's soldiers still crawled
into camp, buried their knives

into his buddies' sleeping bodies.
His wife once painted her kitchen red orange

like the rising sun. The old man

refused to speak to her

until she buried the enemy's shade
beneath a hue of muddy harvest gold.

These were the men we worshipped
in backyards when we played war,

when we stormed beaches with sticks and hunted
hidden enemies deep inside the caves

of our imaginations, before we understood
what must be survived.

Field Dressing

The silent doe stiffened in her bed of leaves,
where moments ago she fell, panting.

Her last breath rattled.
Life passed from nutbrown eyes
into damp January morning.

The snow wrapped us in a womb of silence.

My frost-tipped fingers gripped tight
around the stained handle of a buck-knife,
the curved edge trembling.

Warm against my back, my father's hand.
Soft against my ear, my father's voice—

Careful, son. Cut gently.
We eat what we kill.
We honor the animal.

We honor the dead who give us life.
Boys like me are not made with words enough for this.

Everything Plastic is Melting

When I was a boy,
I thought it would be cool to hang
an empty milk jug from a tree

and set it on fire. Dad said no,
but someone told me the dripping plastic
would whistle like fireworks as it fell.

The neighbor boy, not yet an Eagle Scout,
saw danger in the flames, struck the jug down
with a bat. The flaming container landed

on my head. Viscous plastic melted
into my hands like magma, cooling
into hard clumps all over my hair.

I still have the scars where hardened resin rests
beneath my skin. I, too, contain something
synthetic—a candy wrapper glinting

among the mound of leaves by the banks
of a little stream, or waxy plastic
molecules inside tiny fish.

That night at the kitchen table,
when I hid my wounds with long sleeves
and a baseball cap, my father saw me

for what I was—a blistered imposter,
an exile from the obedient.

Hunting the Ghosts

In mud-soaked jeans, we crouched
into wet leaves, no more than a few yards
inside the wood line. Near enough
to throw the stick and reach our target.
Far enough to conceal our bodies
behind the naked trees and brush.

The empty road below wrapped around
the edge of Porter's Bluff in a hairpin curve.
One twitch of a steering wheel could spin a car
into the dented guardrail and tumbling
hood over roof to the bottom
of our imaginations.

We hunched our bellies tight to the earth.
We saw our grandfathers burning
their enemies out from the caves
of some unnamed Pacific island,
waiting to greet them with bayonets.

We never said it out loud,
but we wanted their terror
for ourselves. We wanted to know
the taste of fear in our throats.

We came here, to the edge
of the woods beside the bluff
to hear the slap of stickpoint
against quarter panel, to run
and hide, to live and tell
our story around the campfire.

BEN WEAKLEY

After the slate-grey Buick rounded the corner
and locked its brakes,
it was our ribbon red and royal blue sweaters
that gave us away.

Hiroshima Dome

after a photo by Michael John Grist

The way the vault arches—
      the way the dome's silent ribs beg

the muted sky for the mercy
      time will not deliver—

bleached bones in the desert.

The windows' vacant stare
      passing through skeletal trees like wind—

sunken eyes staring at the sun.

Something in the way these walls endure—
      how the weathered brick fades—

like the flash
      that turned people into shadows.

My grandfather spent 1945 on Guam.
      He always said the Nagasaki bomb
          was his favorite birthday present.

But the end of the world began here—
      above these walls—

when creation split in two
      and the stars broke into pieces

that our hands can never mend.

BEN WEAKLEY

Triolet on a Line
From a Vietnam War Documentary

In a storm of bullets, in bombs falling like rain,
there was just a village by the river.
A memory passes his window on the train
like a storm of bullets—bombs falling like rain,
children wailing, elders, women gaping at flames.
Thatched huts burnt to cinders and ash. Hearts turned bitter.
In a storm of bullets, where bombs fell like rain,
we were just a village by the river.

America Calls Him

A boy is born
in Reagan's America
screaming Springsteen's
"Born in the USA"
from the bottom
of his purple lungs.

He grows up watching
the action—*Predator*,
Rambo and *Rocky*,
The Terminator.

Remember *Die Hard*? At night,
he shuts his eyes and prays
to the god of manhood, imagines
himself swinging on a rope
from Nakatoni Tower,
screaming *Yipee Kay Yay,*
Motherfucker!

Years later, the towers are punctured.
On CNN, they cough up copy paper
and smoke and secretaries and printers
and insurance company executives
soaked in jet fuel and flame
a thousand feet above the earth.

America calls him that day, young as he is,
to defend freedom. America calls him
to fight *them* over *there*, so that you
don't have to fight them here. America

calls him to bring freedom
to the four corners of the world
so this will never happen again.

In the many years to come,
he will do what is asked
in America's name.
He will do what is asked
in your American name.

Can you forgive him?
Can you forgive yourself?

Soldier's Song

More life exists in the tip
of a bullet smacking
the concrete wall
beside your head

than in a decade spent
commuting to work in traffic
paying the mortgage on time
loving one woman and two children
and taking vacations at the beach.

More sweat pours
more breath gasps and heaves
more heartbeats pulsate
in the intimate space

between shock wave and steel debris
than you could ever find
in any lover's fingers.

More enlightenment
flows through the dust cloud
rising from broken asphalt
to drag you underneath
the opened ground

than in a hundred years
of anything else in this life
or the next.

BEN WEAKLEY

Those who have come close enough
to kiss the painted mask of death—

they have seen millennia.

Pressure

Good Friday, Udairi Range Complex, Kuwait

The first time I saw the sun
rise over the desert
it was 4 a.m.

Across miles of sand
and rusted hulks, the throbbing
of heavy guns.

Over the horizon,
where the beginning and the end
meet and disappear, Friday arrived.

I saw the jeering crowds, the scourge
and spear-tip, the crown of thorns
and the crucifix, waiting.

What could I have known about atonement?
What did I know, then, to judge
the quick against the dead?

There are Four Ways to Die in an Explosion

First the blast rips limbs
from the torso. Throws tender bodies
against concrete walls. Pulverizes
bones against pavement. Those closest
to the bomb are never found
whole.

Then the fragmentation.
Little pieces of metal debris,
like the one that punched
an acorn-sized hole through the back
of Sergeant Carver's skull.

Heat from the explosion starts fires.
Vehicles Burn. Ammunition
burns. People burn
alive. When a driver is trapped inside
white-hot steel, prayers
must be said silently for the smoke
to take him first.

Pressure collapses
lungs and bowels. The bleeding
happens on the inside.
It can be hours
before the skin turns pale
and the bulk of a person
drops.

None of the anatomy is safe.

When the time comes, pray for the blast
or fragmentation. Pray for the heat that vaporizes.
Pray for the kind of pressure
that makes the world dark and silent
before the bitter taste of iron
and cold panic.

BEN WEAKLEY

What is a body to a bomb?

Cathedral of cavities,
empty house of worship
waiting for invocation,
waiting for detonation, blast,
rush of air, the final release of prayer.

What is a body to a bomb?

Vessel. Warm sac of skin,
container of oscillating waves,
made of blood-drenched
and hormone-rich flesh,
collection of tissues
that bruise and burst,
that pulse and breathe
and smile—until they don't.

What is a body to a bomb?

Mannequin
for the clothing of death.
Vehicle
for pipes and fertilizer
stitched into a vest.

What is a body to a bomb?

Audience for a detonation.
Witness to fire-clap, sound-shock,
smell of warmth and rust.

What is a bomb to a body?

Animal skin stretched across a drum.
A hunter's bowstring drawn tight
against the arrow's nock.
A cello string pulled taught, tuned
to vibrate at the violent finger's pluck.

Sunday Morning: The Market in East Rashid

What are we to make of the single brown horse
carrying a slim and neat-bearded rider
bareback, his thighs gripping the animal's flanks
firmly, sandals slapping his heels with every stride,
as the horse gallops through the dust
of a market at midday?

The stalls burst with baskets of figs and dates,
the smell of kebab lamb and warm naan.
Beside the road runs the canal
where we found the Sunni cab driver
beneath a tattered blanket,
swollen, skin-flayed and skull-hammered.

Gray-matter the color of concrete spilled
across the ground, the ribs splayed open
as we interrupted a pack of stray dogs
consumed by their gnawing.

I think of the other horse
we found here the summer before,
bursting with maggots and blowflies,
and I am lost
on the other side of memory
trying to hold on to something that has no shape

when today's horse and rider disappear
into an alley behind the stalls
and the ground around us erupts
when the rockets begin to land.

Sunday Afternoon: Somewhere in Baghdad

The market stalls burst with baskets of figs and dates,
the smell of lamb kebab and warm naan.
A surge of sweat cracks the air when electric
voices rise in the street. There, a crater
remains where a car bomb, dressed up like a bread truck,
parked last month and hit the detonator.
Here the swelling crowd of bodies undulates
while children play around stacks of melons.

BEN WEAKLEY

Checkpoint

East Rashid District of Baghdad,
near the Doura Neighborhood

The car came from nowhere, it came
from everywhere—

white blur and tire squall,
a four-door payload
of heat and pressure and steel.

When it is over, there is just
the tinkle of falling brass and a man
slumped
in a pool of broken glass
and coolant on hot asphalt.

Doc cuts his shirt.
His face is weathered by years
of this. Layers
of skin and yellow fat pucker
from his open side.

He breathes.

In the trunk of the rusted-out sedan,
where the bomb
should be

there are only two tanks,
an oxygen mask, and a box
filled with apricots and dates.

Urgent, Surgical

This is not a poem
in the shape of a 9-line
medical evacuation request.

This song will not be sung
to the tune of radio crackle
and squelch. There will be

no jargon, like *2 PAX—*
LITTER—
URGENT—SURGICAL —
human names will not be translated

into battle-roster numbers
(WHISKEY—SEVEN—SEVEN—TWO—TREE)
for the combat surgical hospital

where the trauma surgeons wait
like sterile deities
in digital camouflage scrubs

for the caved-in skull
and pulverized jawbone
that lies in the staff sergeant's lap

in the bench seat of a Stryker
speeding to the Green Zone
on the stickiest of Baghdad nights

because there is no place to land a helicopter
among the apartments and alleyways

and muffler shops on Market Street

and because the mission disappeared
in the shower of sparks and metal
scattered in the moment

when two artillery shells hidden in a bag
exploded, because the FM radios
could not reach squadron anyway,

and because the voice in the lieutenant's
headset said, without breath, M's
hit—he's got brain leaking out his ear,
which is why instinct told the lieutenant
Go—get the fuck out the kill zone
                    NOW.
Of course, there will be discipline later.

The inquiry into the whereabouts of night
vision goggles that shattered when the molten copper
slug struck the forehead of the combat vehicle

crew member's helmet, called a CVC. Questions
about the route chosen for this patrol. Whether
intelligence briefs were sought or given. Retraining

on how to make a proper 9-line medical evacuation
request before making the decision to seek treatment
for the wounded. Measures to ensure our mistakes

never happen again.

A True War Story

"You can tell a true war story if it embarrasses you. If you don't care for
obscenity, you don't care for the truth; if you don't care for the truth,
watch how you vote. Send guys to war, they come home talking dirty."
—Tim O'Brien, "How to Tell a True War Story"

So there I was ...
          No Shit!

Baqubah '07. A real bloodbath, man.
Bombs under the asphalt. Snipers everywhere.
Everywhere, guns waiting to pump bullets
into fat and meat and bone every moment of the day.

But that's not a war story. This is a war story—

a patrol of armored monsters turns the corner
as dusk wraps the city in death's blanket
and all the lights are out in all the buildings
except the only window

into the only lamp-lit room
in the cracked and crumbling apartment complex
where we see them—lovers, bare-skinned
on a bare mattress surrounded by grey walls.

Check out 3 o'clock—Hajj is fuckin' his ol' lady!
someone hollers into the platoon radio.

Across the city a booby-trapped house
collapses on a squad of Iraqi Army soldiers.
A Hellfire missile hits

a family of seven in their living room.

A quarter mile to the East
a man lies face down in the street,
his head spilled out all over the asphalt
like a burst melon ...

and here we are—
staring through a thermal gunsight, watching
two dark bodies writhing in the heat
at the end of the world.

Heat and Pressure

I.

Awake to the slanted light
on a March afternoon
in Baghdad, I rise
from sweat dried into crust,
salt stained in the shape
of a body on a poncho liner.

The communications sergeant died today.

Knelt down beside a garbage pile
to wipe his forehead,
to rest the weight
of armor and ammunition
on his bones before

the fire clap and dust cloud.
An echoed thump,
felt not heard,
inside a steel tomb.

The urgent voices,
radio squawk and chatter.

send doc quick he's fucked up bad

Punctured tire. Spalled armor.

patriot x-ray ... patriot x-ray
this is sapper five
medevac request follows
litter , urgent , surgical

II.

Sergeant P says
you cannot place blame
when someone is killed
because war is about killing.

Sergeant O says it's best
not to imagine
trading places.

Sergeant J says he didn't hurt.

he didn't feel ... anything
just heat and pressure
and ... like that—

Snap.

everything went black

Night Patrol

Another moonless night. Another flat
cement landscape washed in monochrome green.
We scan the city with night-vision soda straws.
Notice garbage piles, sewage in the street. The fat,
bloated belly of a dead horse. Palm trees
in the concrete. From silence we cross
open roads in heavy bounds. Each boot slaps
asphalt with the weight of knowing. Fragment. Blast wound.
We move through sweat-soaked air, thick as dirt, scratch
our way through darkness like moles tunneling under lawns,
while ghosts watch us from dim-lit rooms above
where their shadows flicker in the windows. Below, we catch
glowing eyes and brace—tremor, shock wave, a bomb's
breath. No—dogs hunting for scraps.

Epitaph for a Deployment

If you could see the crumpled, head-split mass
that used to be a boy slumped down in the street,
if you found him, beside the hole, blasting cap
in hand, the bomb beside his feet,

if you had felt the sweating palms, the ache
of killing, the shame of making godless
what was sacred, you wouldn't buy my beer or shake
my hand. You wouldn't thank me for my service.

Blast

The Burden

She is terrified of the young mouse
stuck in the drip-pan beneath the truck, struggling
and squealing into the poison, drowning
in oil and antifreeze.

She is more terrified of the snakes
that follow young mice into our garage
from the tall grass outside.

I cover the squirming rodent with an old rag
and raise the rubber mallet overhead—
long after I promised God

I'd never kill
another living thing,
I whisper,

Forgive me.

BEN WEAKLEY

Musar Afghanistan
Beside a Dirt Road Cut Into Mountain

An old man appears with hazel eyes
sunken behind the furrowed skin of an almond face
marked by years of sun, work, and death.

I find him in a village, shuffling by the road,
a sickle dangling from his hand. He drags the long tail
of his kameez through the dust while he watches
our caravan of steel cages climb the mountain to Musa Khel.

The only white people he's ever seen
were the Russians who disappeared from the earth
30 years before. No matter what we say
he believes we are Russians.

He knows nothing about the camps
in the eastern mountains, nothing about the foreigners
who lived there. He can tell me nothing
about towers or twisted I-beams, nothing

about airplanes or jet-fuel
blooming into an orange rose
of flame against a translucent sky.

He is still there, in the village cut into rock
below Musa Khel, where we left him
ten years ago. I am gone but he remains.

A Roadside Bombing in Three Acts
(But for the Grace of God)

I.

Around the bend and up the road, just past
the sheer face of wind-carved rock, a dust
cloud rises in a pillar at the blast.
The getaway car bounces by our patrol,
the driver's twisted face of ash
a fleeting ghost. We are early, must have rushed
the bomber into a startled gaffe.
One never knows how or why these things unfold.

II.

For hours we pick up pieces of the man
from the wadi below—a ring, a scalp,
little bits of skull and other things that made
a person hours ago—we throw them in a bag.

III.

We cannot know they were a team of two
until we find the foot and a second left shoe.

BEN WEAKLEY

Getting Left of the Boom

We are taught to hunt them like whitetail deer.
We study their habits, match their wanderings
to the cycle of the moon. We know what makes them
bed down and hide. We know what makes them cautious.

We know they like to feed on the darkest of nights
so we break their routine with helicopters
and low-flying drones, the big white blimp
that floats high enough to see their thoughts.

We deny their animal instincts
until they salivate, then we send out the bait.

Musar Afghanistan walks beside me in the wadi at dawn.
We find it like a static display in the museum of war:

a neat circle cut into hard-packed dirt,
one yellow plastic bucket primed to explode,
a spool of copper wire, a pick and a spade,
and one boy's skinny body, face down,
his dried blood crusting the earth.

Musar Afghanistan shakes his head.
There will be justice, he says. *There will be revenge.*
We must honor him—nothing is over
until the dead have spoken.

No Take-Backs

Musar Afghanistan rides in across a pristine sky
like some bare-chested western god throwing
bolts of lightning, except this time he's hurling

leftover rockets from Soviet days
until he slams one into a dusty path
on Forward Operating Base Salerno
and it bursts into a thousand fragments
where a 25-year-old lieutenant—
the guy who runs the motor pool,

who spends his days hunched over spreadsheets
ordering spare parts, who has not fired his weapon,
who left at home his baby boy—is walking.

He arrives at the hospital in the back
of a dirty white pickup truck
and the big voice on the FOB calls for blood

Type O-Positive and the soldiers line up
around the aid station. A dozen units
and he still bleeds out. It is Sunday.

He'd been taking his dirty clothes to the laundry.

I look at Musar Afghanistan and his smug grin
pressing the wrinkles into his eyes,
and I say, *What the fuck was that about?*

Musar Afghanistan chuckles at this silly game
we've played for years. He never forgets

the score and there are no take backs.

He says, *Gotcha that time, fucker.*

The Orphan of Dowmanda

Wagay Afghanistan is a boy with hollow black eyes
carved into hunger's taut cheekbones,
skin the color of ash, bare feet
the color of the winter mountains around him.

He stands outside the District Center
in a place called *Dowmanda*
with the gaze of an emaciated wolf.

When we approach, he bares sharp teeth
and meets us with a howl: *Sahib, sahib—candy!*
an act he learned begging at the gates
of the outpost named *Wilderness*.

We tell him to come back tomorrow,
because tomorrow we will not return
and today we have nothing for him.

Musar Afghanistan
Speaks About the War (I)

The old man, Musar Afghanistan sits
on the corner of a sweat-stained mattress
in a mud-brick hut, beneath a roof
thatched together with clay and straw.

He combs his henna-stained beard
with leather-padded fingers and yellow nails
while the quiet boy fetches hot water
for tea. I ask him why the young men attack
always from his village, why his young men
put bombs in the ground.

Musar Afghanistan says,
                    Who knows
the minds of young men—you pass the time
in your way, they pass time in theirs,
and I will be here in my garden, waiting.

Musar Afghanistan
Speaks About the War (II)

God is the drone that circles above.
God whispers in the warhead falling from cloud and sky.

Pain is the finger that squeezes the trigger.

Sweat-stained gloves silence the grip and numb
the skin. The shooter feels nothing.

Let me tell you about sacrifice, the ritual
pouring of blood

at the edge of a knife. Have you seen the devil?
Let me tell you about the devils I see

in that pillar of fire and cloud
that swallows the world into an empty nothing.

Nothing remains but dust. Let me tell you
about the dust

that holds my father's blood.
How I pressed my palm into the crust,

how I squeezed the soil
between my fingers, how I felt the generations.

You may leave, but the devil always returns.
Next time, I'll be waiting,

one hand full of the earth, the other
gripping the knife.

For Those I Love, I Will Sacrifice

Pried loose from the rock that bound him to earth,
six-thousand feet above the sea, he was pulled
from red-tracer dreams of love and womb.

Tonight, when there is nothing left
but darkness and breath,
let him sing.

Stay. He will sing for you of his memory made viscous
by the years. He will sing of his valley, where ghosts
disappeared into mountain haze.

Here, where we cannot look away, let him sing.
Hear the song of ball bearings and fertilizer,
batteries and copper wire.

Let him sing. Hear him until you know heat and pressure.
Hear him until you feel blast wave
and ruptured lung.

Listen. He will sing of the golden hour,
songs of pale skin and translucent bag.
Tourniquet and rotor wash.

Let him sing. Hear the desperate music of splint and bone, hovering
over the valley floor.

Let him sing to you of what he left in the mountains
so that, once, you can bear the weight of his body
armor. You, too, can hold the souls he carried.

Let him sing of sunken eyes and rusted rifles.

Let him sing of dark rooms and phantom limbs.
Faces he sees in broken mirrors. Ghosts
he does not recognize. Ghosts that look like his friends.

Let him sing what it is to touch their faces in his dreams,
so that you, too, can wake twisted in sweat-soaked sheets.

Let him sing so that we may feel
his voice, because we must feel his voice.
Not as wind. Not as the moment
of breath against our skin, but in the ritual ache
of our memory, where he belongs forever.

Debris

Ode to a Scar

Pale pink skin,
one inch long
and one-and-a-half wide,
sunken into the gutter
beside the tibia.

A jagged bowl of hairless flesh.

The war-wound began as a scrape
when I fell into an open sewer
hole beside an elementary school
in Baghdad. I walked right past
the glowing chemical light sticks
that were placed just-so on the ground
to prevent this moment—

the lieutenant, charging ahead
into the two-dimensional tunnel
of monochrome green
in his night vision goggles
before disappearing, waist-deep
in a real shithole, hanging
by plates of body armor.

The soldiers of second platoon
raced to the phones, sacrificing
hot chow and warm showers after mission
to be the first to call home, laughing.

Day after day, the scrape oozed
red-brown into my pants leg

during November's long patrols.
I carried the dirt of that place
beneath my skin for months.
Antibiotics did nothing. All winter
we bathed in flames,
the scent of burning tires
soaked in jet fuel. We breathed
the black fumes of burning vegetation,
clearing fields of fire
for the infantry. The particles
entered our pores, our eyes,
our tongues, the tiny air sacs
in the deep of our lungs, pieces
of war that travelled home inside us
like dust tucked in the creases
of old uniforms, like the infection
that lived in my leg
for so long after we returned.

My wound is healed now.
Another thing I carried home.
But it is not gone.
When I wear shorts, you can see
the dark-ringed scar above my socks.
Not the original skin, but flesh
transformed. Still there but hardened.
A memory no longer open or weeping.

On Douglas Lake

Water is a healing thing. Cloud is a healing thing.
The mountains' cold blues and greys surround us.
They, too, are healing things.

There is no war here, no shower of sparks.
There is no recoil here, no dust plume, no concrete spall
where rocket meets bunker. No bullet
will burn this sky with its red-tracer arc.

Three clusters of geese burst low in a line
across the water, honking their return.
The crescent sandbar rises to meet them.

Emerald and auburn hills
brush across my eyes and my eyes devour their texture.
English Mountain feeds my eyes
with ridges and folds curled like knuckles.

Banded layers of rock rise, exposed
at the border of the low lake. Limestone,
rippled and heaved by time's heat, grazes
the skin of my arms and my chest—
I love its hardness.

My fingers trace the line of these mountains.
My skin presses against the ridges and valleys,
rising and falling like breathing bellies.

In the morning, the blue-orange mists
dip low beside calm waters. The sun rises
to part the haze and say, welcome to this place.
Welcome to your home that always was.

BEN WEAKLEY

Omen

An old, black crow
rests on the top rail

of a wrought-iron fence,
keeping watch

over garbage
cans in summer morning mist

split by daybreak sun.
Last time, old friend,

you clotted the skies
over Baghdad, Ameriyah,

Little Market Street—
a terrible clap of fire,

dust plume, heat
and pressure,

steel fragments.
What if

the dust of that day
became ash, reached cloud

and crossed oceans
in the jet stream to find me?

What, then, old friend,

have you come to tell me?

What, can you say,
old crow,

about the living
or the dead?

Where It Lives

The fear still lives in my skin
where electricity dances down nerves
and across synapses
whenever the dog bares its teeth
and snaps at my hand
at night when I pick him up from our bed.
Or when I am driving home in the dark,
our Monday night routine,
the boy in the back taunting
his sister, the ballerina,
with his latest martial arts technique.
When she screams,
the way children scream when sibling-struck—
I yell
   God
     dammit

get the fuck down!

   Get Doc here—now!

and the car is silent
except for the clicking blinkers
and the rush inside my head
where reflex takes me
to muscle memory,

       where churning gut

and trembling hands live,
where lifetimes are exhausted

in the scent of sweat and diesel,
cordite burning, garbage burning,
and concrete dust billows into cloud,

and the rust-shit smell of death

smothers the living world. When I pull
the car into our garage
the children are silent, remain
silent, because by now they know
their tears will only take me further

into the world my body cannot leave.

I wake to drowning

in air. I sweat my salt
to the midnight moon.

Each night I put my head down
and cover this body
in high thread count armor.

I fight God
and the Devil in my dreams
until—waist-deep

in dog-tags and empty brass—
I run out of friends and ammunition.

Pantoum for Watching the Syrian War on Television

Before dawn in Ghouta, the rockets fell
like fists from an angry father. Steel broke the ground
open, burst into concrete walls that held
the sleep of families. Then came poison clouds

like fists from an angry father into broken ground
where a thousand wet lungs seized and coughed.
The sleep of families broken by poison clouds.
The bodies laid in rows with their arms crossed.

A thousand wet lungs seized and coughed
through tremors and foamy mouths.
The bodies laid in rows with their arms crossed,
their blue lips parted and eyes bulging out.

I see tremors and foamy mouths,
pale children stiff in stained pajamas.
Blue lips parted and eyes bulging out.
The news cycle repeats these traumas:

the pale children, stiff in sweat-stained pajamas,
Red Lines, Barrel Bombs, words of war,
the news cycle repeats these traumas
between ads for cold beer, discount furniture stores.

The Red Lines and Barrel Bombs—words of war
echo against the mirror where I shave
while ads for cold beer and discount furniture stores
distract from death six-thousand miles away.

Echoing in the mirror where I shave
is today's list—grocery run, overdue bills—
distractions from death six-thousand miles away.
I am safe here. I have prescription pills.

Today's list. Grocery run and overdue bills.
I see bombs bursting concrete walls that held—
I am safe here. I have prescription pills.
Today in Ghouta, the rockets fell.

How Will You Answer

What is the word for *home*
after houses become bombs
as they did in Baqubah and Mosul?

One afternoon your wife
has you drill pilot holes
to hang a flat-screen TV on the brick wall.
The mortar dust and shards of clay

erupt from the spinning bit
like bone ejected from kneecap
and skull in the Baghdad torture rooms.

At night, you put your son into bed
and draw the blankets up
over his freckled shoulders.

You stroke his straw-blonde hair
and wonder, what
is the word for *son*, now?

What can you call your son
now that you've seen another man's son
burning?

How will you answer
when your son calls you *father*
in the world you turned
into ash and bone?

BEN WEAKLEY

Anything that Can Happen

Murphy's Law states:
anything that can happen
will happen. If the physicists

are right, then somewhere
in the giant loaf of spacetime
my son sits on the concrete floor

surrounded by blood and broken glass
holding his wrist, skin parted to bone,
warm red jets pumping,

pumping, pumping, across the room
while his screaming sister
hunts for old towels.

I imagine his face whitening,
draining, his purple lips shivering,
as his mind, his ten-year-old mind,

recognizes that it has run out of time.
I imagine his mother screaming
at the 9-1-1 operator, who cops

attitude because my wife admits
she doesn't know how to prepare
a tourniquet. I imagine his words

I'm sorry momma,
it was an accident,
I didn't mean to—

I imagine him taking responsibility

for his death because even at ten
he is always taking responsibility
when it isn't his fault.

But here, in the part of the universe I can see,
the boy sits in the front seat of the Toyota,
a towel pressed firmly into a wound
so deep you can see tendon and bone

and vessels intact. The bleeding stopped
before my wife called—her first words:
Don't panic now, everyone's okay.
The boy tells me it doesn't hurt,

says he is not scared, his voice calm
over the line as I gather my things
to leave work.
My wife sends me a photo captioned, *After,*

and five stitches is all it took to travel
halfway around the boy's arm.
When she sends *Before*, I see skin flayed
to muscle and bone, the way I have not seen

that deep inside a body in ten years—
and I am lost now, staring into memory's
blood-grimed face. I feel hot breath
against my neck, holding the scent of earth

and sweat and pomegranate.
Musar Afghanistan smiles,
pulling wisdom from his slick and purple beard.
I knew you would come back to me, he says.

You always come back to me, old friend.

59

Fragmentation

Ghosts at the Border

after a painting by Denise Shaw

Among the grey-green shadows
in the green-washed aperture
rise the spectral white-green wisps
of humanity, framed
by the crosshairs of a night
vision scope. They are alive
inside numbered reticles,
mists congealing into hands
and feet, legs and arms, children
with terror-twisted faces
and ethereal torsos.

What will we do with them now,
these phantoms and their white-hot
flesh, dancing in the spotlight
toward the border patrol trucks
where the agents are waiting
with zip-ties to take them
to the fetid cages?
What will we do with them now
that we can recognize this
mother's face, painted thick
with resolve, unmoved by fear,
illuminated by the faint light
from stars we no longer see?

BEN WEAKLEY

In Some Distant Country

We have seen this before, in books
and on the screen, like dust plumes rising
in some distant country. Except,
some distant country is Michigan—
armed patriots (terrorists)
in the marble halls of a statehouse.
Long guns and body armor.
Stars and bars on the flags they carry
and nooses for the nervous traitors (lawmakers)
who can read the signs on the lawn outside—
TYRANTS GET THE ROPE.

Now they are here, inside
the United States Capitol Building,
these armed patriots (terrorists)
smearing their urine and their fecal matter
on the floor and the walls, roaming
the halls with zip ties and body armor,
looking for traitors (lawmakers)
to bind, to carry outside,
where the gallows wait.

Their work is not finished.
Tomorrow, these armed patriots (terrorists)
will return to their homes, victorious,
triumphant. They will return
to towns across the fifty states
where they work at hospitals and gas stations,
at schools and police stations. They will smile
when they greet us in the grocery store
while they do their shopping.

They will tell us to unite.
They will tell us to listen
and be calm, that time
will grant amnesty (without repentance).
They want us to forget, but
their work is not finished.

Who will tell us how to love
our neighbors now?

Who can show us how to rescue
our would-be executioners
from the gallows they built?

BEN WEAKLEY

As I Watch the Capitol

I feel it in the crook of my clenched jaw,
the tension of teeth pressed into teeth,
the raw gum, the incessant ache
of bone sliding past bone along the socket.

Mouth, I will call you *fury*.
Mandible, I will name you *indignation*.

I will not call your pipe bombs
patriotism.

I will not name your zip-ties
freedom.

The war that you are asking for,
the civil war that you make impossible
to avoid, it has many names:

it is rotting flesh peeling from stiff bodies in the sun,
is is torture rooms with power tools and chains,
it is dried blood stains against damp cinder-block walls.

It is a bride in the back of a pickup truck—
her head bouncing against the tailgate,
spilling grey matter and long black hair

from a hole made wide by shrapnel
like a suicide bomber's eyes
before the blast.

Her name is *misery*.

May you turn away
before we all must meet her
and her siblings *agony* and *regret*.

BEN WEAKLEY

Musar Afghanistan in Rawlins Park

I am walking through Foggy Bottom
down E Street
toward the parking garage

when I see Musar Afghanistan,
the old man sitting on a bench in Rawlins Park
where the cherry blossoms are in peak bloom

bursting their pinks and whites
around the pond and the fountain
and blocking out the district's gray facade

and I sit down beside him, Afghanistan,
with his trimmed beard and polished nails
and his silver hair. Now he wears

boat shoes and Ray-Bans and a sharp blazer
while he eats kebab from a food truck
beneath the weathered bronze statue

of a forgotten Union Army general.
I sit beside him on the bench
in Rawlins Park, beneath the weathered bronze

statue, surrounded by the cherry blossoms
shielding us from the heat that oozes
from the district's grey façade,
and I ask him what he's doing here,

on a bench beneath a bronze statue
in the middle of the cherry blossoms

bursting their pinks and whites
around the pond and the fountain spraying,

I ask him, Musar Afghanistan, if he is real,
which is to ask if he really exists
here, in this place outside memory,

and Musar Afghanistan wipes his mouth
with a paper napkin and says:

I am here because you are here.
Wherever you go, I will follow.
War is not finished with us, friend.
War has made us more than brothers

Three Weeks Later

and just like that, life is back
to the way it was the day before
a grown man prowled the marble halls,
growling and shirtless, his runic tattoos
exposed beneath a bearded face and horned
Viking helmet. We see him now
in memes on our phone, harmless
as a drunken football fan,
or one of those make-believe Nordic raiders
from a Capital One commercial
where the barbarians tour the Grand Canyon
or stroll through the aisles of a grocery store.

The voices from my screens surround me,
saying,
      unify, unify
and we return to work on Monday and worship
at church on Sunday and no one says a thing
about the gallows, or the noose,
or the pipe bombs,
or the elected officials who told us
that night, January 6th,
                  count me out—enough is enough
and
      this has gone too far!

Now the voices on my screens, surrounding me,
shout
      Unify! Unify!
       You are the extremist, you
        are the one tearing us apart!

Was what happened that day
some smoke-obscured dream? Am I alone
believing that I saw the red, white, and blue
sea of humans crushing the officer's body
in the jaws of a swinging door, beating
a helpless head with a fire extinguisher?

I no longer trust myself to know
what I know and the screens are everywhere,
whispering —

> *unify, forget.*
> *We can put this all behind us.*
> *Unify, forget.*

BEN WEAKLEY

How to Forget a War Crime

Wear matching pajamas on Christmas morning
when you watch your children open their presents.

When you watch your children, remain present.
Try not to remember the traffic circle.

Do not think about the traffic circle
when you adjust the seat on your son's new bike.

Take pictures. Capture this. Your son on his bike.
Nisour Square. Tire-smoke. Burning sedan.

Don't think of Nisour Square, the burnt-out sedan,
the shock-eyed faces of death, their gaping mouths.

Push away these dead faces, their gaping mouths.
They were *Iraqis*, no Americans died.

Tell yourself no Americans died that day.
Wear matching pajamas on Christmas morning.

A Soldier's Lament

And who remains to listen
when we tell our stories?

In the land without memory,
In the country of words
that have no meaning,

who remains to hear us
when we speak of our war?

BEN WEAKLEY

Life is About Choices

Life is about choices—what we'll hold, what we'll lose.
The dead remain forever the dead,
but the living are alive if they choose.

Long winter shadows lengthen winter blues
and rows of barren trees bind the road ahead.
Life is about choices—what to keep, what to lose.

Someday I'll become dust. Still, the trees will bloom
white petals and green leaves that turn fall's brown and red —
the living live just a moment if they choose.

I drive for miles, absorbing the news.
This world steeps in hate and wallows in dread.
Life is about choices—what's kept close we abuse.

These fears I squeeze, they choke me like a noose
while continents burn, war drags on and spreads —
we're condemned to this moment, whatever we choose.

What joys I hold slip away like an afternoon
sky bowing to the moon, its truth unsaid.
Life is about choices—what we hold, we lose,
but the living can live for a while, if they choose.

Once It Is Over

When the nation
has been served,

when there are younger,
better men, women,
children
to fight our wars,

you move on
to enjoy your twilight years,
your remaining years

(oh god, you are only thirty-eight
and everyone says
that is young, so young).

Once you are pronounced
non-mission capable,
no longer serviceable, disposable,

when there are no parts available
to repair the machine
because they no longer make
those kinds of parts,

when you are sent to the depot
for retirement and de-
militarization, this
is what remains—

the pain,

BEN WEAKLEY

the ache
like a dark and purple crust,

like the black blood pooling and pressing
against the surface
of your swollen skin.

What Remains

What remains are the scars,
blackened patches of earth among the grass
where the campfires were lit.

We hold the memory against our skin—
the night sky, starlight, constellations
in the absence of cloud.

The weight of breath spilled
like fog from our mouths.
It was enough to be near

the open flame, to warm ourselves
so that we did not succumb
to the frost.

Sometimes it is enough to endure.
Sometimes we must accept
permission to survive

until the dawn brings an orange sky
down against the mist.
What remains is bone, knuckle,

cartilage. The tissues thaw and awaken, plump
and engorged with the fresh blood

that brings pain and warmth
in equal measure.

The Wooden Elephants of Herat

I type *Afghanistan* into a search engine
that spits out words connected to places
and I get more places: *Kandahar, Khowst,
Gardez, Herat.*
      I never deployed to Herat.
But Herat is where a woodcarver cut
scraps of walnut into two elephants
I brought home from the war to give my son.
For eight years they roamed his room as he played
in the ivory carpet of his imagination
until the tusks, tiny as matchsticks, fell out.

He is ten now. He does not remember teething
on my dog tags or holding my sweat-stained
patrol cap in the Fort Knox gym the night I came home.
He does not remember stopping the car
to salute the flag when the trumpet played retreat
on post. He no longer plays with elephants,
and now I pack them into a cardboard box
with faded uniforms and dusty boots —
the relics we're unable to throw out
but no longer want to display.

Acknowledgements

I am grateful to the editors of the following magazines and journals, in which these poems first appeared—some in slightly different versions:

"Checkpoint" first appeared in *The Wrath-Bearing Tree* November 2020

"Field Dressing" first appeared in *Cutleaf Journal* November 2022

"Good Friday: Udairi Range Complex, Kuwait" first appeared in *The Wrath-Bearing Tree* November 2020

"Heat and Pressure" first appeared in the anthology *Proud to Be: Writing by America's Warriors, Vol. 10*, Southeast Missouri State University, Cape Girardeau, Missouri, 2021

"Hiroshima Dome" first appeared in *Ekphrastic Review* May 4, 2019

"How Will You Answer" first appeared in *The Wrath-Bearing Tree* November 2021

"In Some Distant Country" first appeared in *The Wrath-Bearing Tree* November 2021

"Soldier's Song" first appeared in the *Line of Advance* journal, where it won a second-place in poetry in the 2020 Col. Darron L. Wright Memorial Writing Awards.

"Sunday Morning: East Rashid" first appeared in *The Line Literary Review* Spring 2021

"There Are Four Ways To Die in an Explosion" first appeared in *The*

Wrath-Bearing Tree November 2020

"No Take-Backs" first appeared in the *Line of Advance* journal, where it won a first-place in poetry in the 2021 Col. Darron L. Wright Memorial Writing Awards.

"When We Were Boys" first appeared in *Cutleaf Journal* November 2022

"The Wooden Elephants of Herat" first appeared in *Cutleaf Journal* November 2022

"For Those I Love, I Will Sacrifice" received a first-place in the 2019 Heroes' Voices National Veterans' Poetry Contest.

A Few Words of Thanks

Thank you to Seema Reza, the first person to treat me as an artist. Thank you to the Community Building Art Works (CBAW) community for sustaining my faith in the goodness and decency of humanity.

Thank you to Sandra Beasley, Meg Eden, Denton Loving, Lisa Kamolnick, and Howard Carman for reading early drafts of *HEAT + PRESSURE,* and for the encouragement to keep going. Thank you to my publisher, Randy Brown, for helping make it a reality. I wouldn't have wanted to bring this book into the world with anyone else.

Thank you to my fellow writer-veterans, who took time to review and endorse *HEAT + PRESSURE* prior to publication: Brian Turner, author of *Here, Bullet* and *Phantom Noise*; Colin D. Halloran, author of *Shortly Thereafter, American Etiquette*, and *Icarian Flux*; David P. Ervin, author of *Leaving the Wire: An Infantryman's Iraq*; F.S. Blake, author of *Terminal Leave, Above the Gold Fields*, and *The Few Drops Known*; and Christopher Lyke, author of The Chicago East India Company; and Martin Ott, author of *Lessons in Camouflage* and *The Interrogator's Notebook*. I am humbled by their words and insights.

To Fred Johnson, John Flaviano, Shane Grantham, Ron DeLurme, Matt Lashley, Josh Walls, Orlando Craig, Josh Wiles, Joe Frederick, Nick Paolini, and many more wonderful people with whom I've served—you taught me how to soldier and how to lead. I am eternally grateful.

Thanks to Dad, for showing me daily that the future is not limited by the past, and that we all have a second act in us.

Thanks to Stefanie, for bearing all of this with me and for loving me anyway. Thanks to my children, Abby and Jack, for filling my life with joy, for being so delightfully weird, and for inspiring me every day.

Artist's Statement

The impulse for *HEAT + PRESSURE* comes from the same place that first inspired me to start writing poetry: I needed to bear witness to the world. I needed to examine the wars in which I had participated as an officer in the United States Army during the peak years of America's Global War on Terrorism (GWOT).

I began writing when I was still on active-duty. I don't draw. I cannot paint. Don't ask me to sing or play an instrument. Poetry, though, allowed me to render into images that which couldn't always be explained with logic and prose. Poetry allowed me freedom to take on the ephemeral ghosts of war and to affix them meaning on the page.

I wrote these poems on phone screens in the back of commuter buses and metro trains. I wrote them in notebooks on lunch breaks at work and quiet Sunday mornings before the rest of my family was awake. I wrote them on laptops and gas receipts. At first, I wrote about combat, acts of war, and unprocessed violence. As I pushed myself further, I wanted to know more about *why* I had decided to go to war, in a country with a volunteer army. I wanted to know what it meant for my life now. The words happened. I bore witness.

There is nothing special about me. My experiences in Iraq and Afghanistan were mundane and terrifying, touching and lonely, humane and de-humanizing, sometimes all at once. There's nothing unusual in all of that.

Some soldiers experienced much more and much worse. Some experienced less. I share these poems in the hope that other veterans and the people who love them will see something of themselves reflected back.

I also hope readers with little military knowledge or awareness will find and read this book. Being a veteran of war can be isolating and lonely. In the process of writing this book, I was lucky to find community with people who made me feel heard and seen. I now strive to make similar opportunities happen for others.

About the Writer

Ben Weakley writes poetry and non-fiction to bear witness to war, and to the human experience in crisis. In a 14-year active-duty U.S. Army career, the veteran experienced deployments to Iraq and Afghanistan, and one very long tour at a desk inside the Pentagon.

In 2018, as he was retiring from the Army, Weakley began participating in writing workshops with Rockville, Maryland-based Community Building Art Works (CBAW). The practice sparked in him a passion for creating opportunities for dialogue among "civilians" and "military" audiences through writing and literature.

Weakley enjoys creating spaces for people to speak truths and tell their stories, with special care toward those who aren't usually heard. He himself now facilitates on-line writing workshops through CBAW, as well as through the Philadelphia-based non-profit Warrior Writers.

His poetry and non-fiction appears in such literary publications as *Cutleaf Journal, Sequestrum*, and *The Wrath-Bearing Tree*. His work has been anthologized in *We Were Not Alone* (CBAW, 2021) and *Our Best War Stories* (Middle West Press, 2020).

He was the 2019 winner of the Heroes Voices National Poetry Contest and the 2021 poetry winner in the 2021 Col. Darron L. Wright Memorial Writing Awards. The latter is administered annually by the Chicago-based literary journal *Line of Advance*.

Weakley lives in the Tri-Cities of Northeast Tennessee with his wife and two children—and a well-meaning but poorly behaved hound-dog named Camo.

Find him on the web here: **linktr.ee/benweakley**

Did You Enjoy this Book?

Tell your friends and family about it! Post your thoughts via social media sites, like Facebook, Instagram, and Twitter!

You can also share a quick review on websites for other readers, such as Goodreads.com. Or offer a few of your impressions on bookseller websites, such as Amazon.com and BarnesandNoble.com!

Recommend the title to your favorite local library, poetry society or book club, museum gift store, or independent bookstore!

There is nothing more powerful in business of publishing than a shared review or recommendation from a friend. We appreciate your support!

You can write us at:

Middle West Press LLC
P.O. Box 1153
Johnston, Iowa 50131-9420

Or visit: **www.middlewestpress.com**